VOLUME 75

INCORPORATION OF THE SPIRIT INTO THE PHYSICAL BODY

PROBABILITY IS NOT THE QUANTUM VALUE OF THE ENERGY

First Edition

Carlos L Partidas

ISBN: 979 8352 1211 77
SAPI INTELLECTUAL PROPERTY REGISTRATION: N° 8074
OF THE COMPENDIUM THE CHEMISTRY OF DISEASES
BOLIVARIAN REPUBLIC OF VENEZUELA, 07/05/2010

DEDICATION

FOR MILEVA MARIĆ RUZIĆ, THE SERBIAN PHYSICIST AND
MATHEMATICIAN, WHO WAS PROBABLY THE AUTHOR OF THE
THEORY OF RELATIVITY. MILEVA MARIĆ SHOULD BE AWARDED THE
MERIT FOR DEDUCING THE ENERGETIC ECUATION $E=mC^2$. THIS IS
THE EQUATION THAT ESTABLISHES HOW ENERGY IS CONVERTED
INTO MASS, WHICH REVOLUTIONISED HUMAN THINKING. ONLY
THAT MASS REFERS TO PARTICLES THAT HAVE NO MATTER

CONTENTS

RECOGNITIONS

TO THE GERMAN CHEMIST HERMANN EMIL FISCHER. WITH HIS CONTRIBUTION, WE CAN UNDERSTAND THE CHEMISTRY OF LIFE

THE DUTCH CHEMIST JACOBUS HENRICUS VAN 'T HOFF. A HENRICUS VAN 'T HOFF IS RESPONSIBLE FOR THE EXPLANATION OF THE PHENOMENON OF CHIRALITY, WHICH IS FUNDAMENTAL TO THE UNDERSTANDING OF THE THREE-DIMENSIONAL FORM OF MAGNETIC MASS AND ELECTRONIC MATTER

1

PROBABILITY OF ROTATION

The electronic matter of a physical body is different from the magnetic mass of a physical stamp. The electronic matter of a physical body was formed by the integration of electronic energy, whereas the magnetic mass of a physical stamp is the result of the integration of magnetic energy. Electronic matter forms a solid physical body, whereas magnetic mass contains absolutely no electronic matter; therefore, magnetic mass forms a three-dimensional holographic figure that resembles electronic matter; but this magnetic mass is an energy that contains no electronic matter at all.

For example, spirits are made up of magnetic mass without any electronic matter. The weight of electronic matter depends on the point in space where the electronic matter is located; therefore, from a gravitational perspective, each point in space is different; and, at each point in space, we will have to refer to the weight of the electronic matter but not to the magnetic mass; since, magnetic mass is not affected by the gravity of physical space. Weight refers to the accumulation of electronic matter, whereas the attraction of electronic bodies does not influence the magnetic mass of a spirit.

The two definitions are obviously distinguishable; or let's say, electronic matter is physical; therefore, we can see electronic matter from the physical world. Whereas the magnetic mass of a spirit cannot be seen by everyone in the physical world, because magnetic mass is a consolidated state of energy, which can only be seen by people who have an amplified visual range, or who have a visual range similar to the visual range of a baby, or a child up to the age of 5 years old.

Another aspect of electronic matter is that electronic matter is not aware of the existence of itself; therefore, electronic matter is not aware of the existence of the Universe. Whereas magnetic mass is aware of its existence and the existence of the Universe. Magnetic mass is the energy that characterises and moves the electronic matter of a physical body; or it is the energy that defines the duality and quality of existence of a living being.

In Figure 1 it can be seen that, in the beginning, electronic matter was formed from the spontaneous integration of the two electronic energies rotating in the same direction. The integration of these two positive energies can be explained by the probability of the direction of rotation. Energy is produced by motion; therefore, the motion of the energy can be explained by the spin probability. So, the spin of energy can be explained by fermions and bosons. For example, the motion of energy can be explained by fermions, because we don't know what the value of energy is, so the levels are probabilities, but they are not quantum levels.

Fermions that are spinning in the same direction can be integrated without this integration affecting the energy value at that level. For example, 2 fermions spinning from left to right, i.e., +1/2 and +1/2, can be spontaneously integrated. And, the 2 fermions that are spinning in the right-to-left direction, i.e., the -1/2 and -1/2 fermions, will spontaneously integrate. The union of these two positive and negative fermions can be explained by the integration of a BE boson.

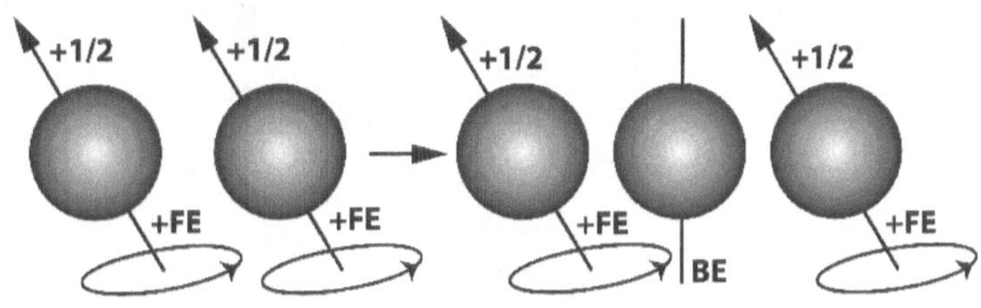

FIGURE 1

**CREATION OF ELECTRONIC AND MAGNETIC ENERGY FB
BY THE MOTION OF AN ALMATRINO**

In Figure 1 we show the 2 fermions pointing upwards. We can see in this Figure 1 that the spin probability of a fermion creates two kinds of energy by the motion: the electronic energy points with an upward energy flow direction; and the motion of this electronic energy creates a magnetic energy that spins perpendicular to the flow of the electronic energy.

The probability of the next fermion to form at that level must necessarily point its energy flow downwards, so that the two fermions can contain the same amount of energy at the same energy level. This amount of energy added together will give us an energy value that will not change the value of the energy at that level of probabilities.

In this case, Figure 2 represents the first energy level from which the Universe began to form. The fermion 1 in Figure 2 is the spin probability of the first particle that connected nothingness to what is now the Universe; therefore, it must have been the smallest particle that the mind of a human being with rational analytical ability can conceive of.

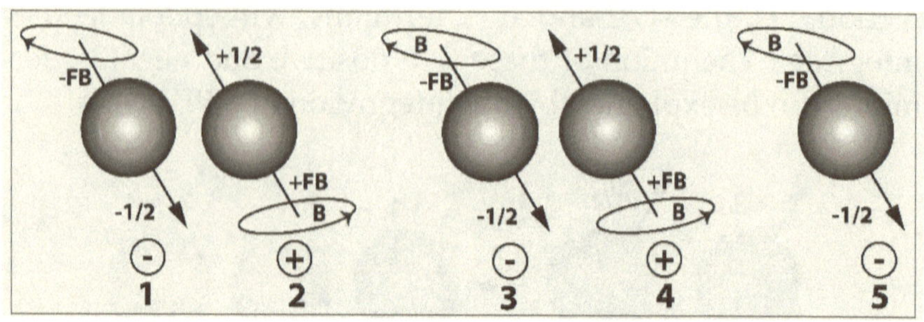

FIGURE 2

FORMATION OF THE FIRST ENERGY LEVEL OF THE UNIVERSE

The value of the amount of energy of the first level we will not know, since we can only analyse the probabilities of the interconnection between the elementary particles.

The spin probability of fermion 5 in Figure 2 is the one that connects level 1 with level 2. Fermion 11 in Figure 3 is the one that integrates fermion 12 with the third energy level.

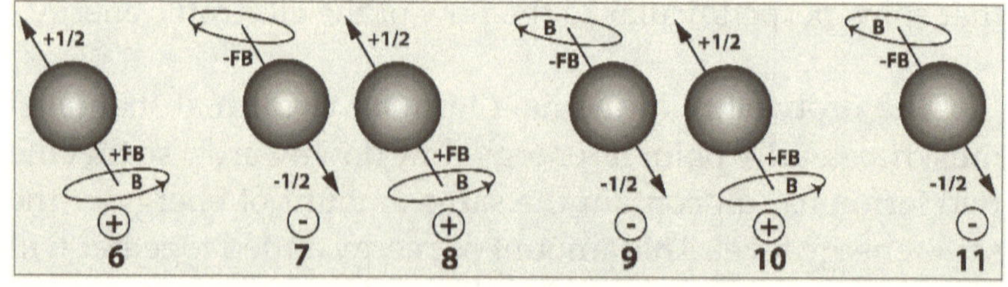

FIGURE 3

**ENERGY LEVEL 2, HAS A PROBABILITY OF
INTEGRATION, WHICH CAN BE REPRESENTED BY A SUCCESSION OF
ENERGY LEVEL EVENTS**

It will be in this sequential way, in a succession of energy levels that we can represent by the interconnection probabilities; and whose sequence goes towards an infinite value of energy levels.

For a second particle to form, it will be a fermion with a spin probability indicating that the flow of its electronic energy will be pointing upwards, i.e., in the opposite direction of particle 1. Particle 3 will have its energy pointing downwards, just like particle 1. Therefore, at each energy level we will have 4 particles. The motion of these 4 particles can be represented by 4 fermions. The 2 fermions 1 and 3 point their energy downwards; that is to say, the fermions 1 and 3 are negative fermions; which integrate spontaneously and form a union that represents the negative electronic matter.

This form of the probabilities of motion indicates that the first particle that formed the Universe spun from right to left; that is, the first particle that formed the Universe was a positive particle whose electronic energy flowed upwards.

Fermions 2 and 4 point their energy upwards, i.e., we can define this energy value as positive. Probability tells us that these 2 positive fermions will integrate spontaneously. And so positive electronic matter is formed.

A boson is a probability that the two fermions, which can be positive or negative, will integrate. So, a boson represents a 100% probability. We cannot assign a direction of rotation to this probability; it is an absolute probability. For example, the sum of the spin probabilities of a boson is two fermions of 50 + 50 %; therefore, this sum does not change the absolute value of the probability at that energy level, which is a probability that will remain 100 %.

That is, in a physical way, a boson is placed between two fermions, whose energy flows in the same direction to integrate them. Fermions can be positive and negative, but the positive and negative sign of a fermion is not a mathematical sign. The sign is only used to indicate the direction of rotation of an integration probability. So, there is no such thing as a boson that has zero energy; that definition has no physical meaning. It is a probability of motion.

This minimal initial particle that generated the energy by the motion, and from which the Universe was formed, we had to define as an almatrino. An almatrino is only energy; in such a way that the energy of an almatrino has neither electronic charge nor electronic matter; nor does it have magnetic mass. In other words, an almatrino is only energy; and this initial energy cannot be defined as magnetic energy or electronic energy. An almatrino is only the initial energy that began to move in that first instant. It is from the initial movement of an almatrino that all the electronic matter and magnetic mass that exists in the Universe up to this moment was generated.

The existence of these two kinds of energy can be explained by the energy equation $Ev=m_0C^3$, which was derived from the equation $E=mC^2$. Let us say that $E=mC^2$ is the energy equation of Mileva Marić, until history clarifies the situation of merit. However, everything indicates that the authorship of this energy equation of the conversion of electronic matter from the integration of electronic energy is due to the Serbian mathematician and physicist Mileva Marić. However, in making this deduction, the fact that magnetic mass is different from electronic matter was not taken into account.

The electronic matter that exists in the Universe up to now was formed by the integration of electronic energy. The electronic energy of the Universe will continue to be formed by the perennial motion of the Universe. The motion of the Universe is and will be perpetual, because the Universe will always be in motion; for, the Universe is expanding against nothingness.

In nothingness nothing exists; therefore, in nothingness there are no forces that oppose the growth of the Universe. Therefore, the movement of the Universe will be progressive and continuous towards nothingness. As there is more motion, more energy will be created; therefore, the growth of the Universe will be eternally towards nothingness. Nothingness is the same size as the Universe, and as the Universe grows, the boundary separating nothingness and the Universe moves away towards nothingness.

Energy is created by motion; therefore, it will not be possible to stop the expansive motion of the Universe.

There are two ways for the Universe to reach a state of thermal equilibrium. One is that the Universe itself is creating space; therefore, as the space created by the Universe itself increases, the Universe dissipates its thermal energy. The other factor that achieves the thermal equilibrium of the Universe is that the Universe increases its entropy; therefore, the Universe is expanding chaotically towards nothingness; since, in nothingness there is nothing that can establish an order.

So, the motion of the Universe creates its own energy; and this is the energy that drives itself towards nothingness; therefore, the motion of the Universe will be perennially towards nothingness.

There is no energy that has no motion, and because it is energy, energy will always be in motion. Someone is not going to come and say that it was a being called God who made the almatrino move in order to create energy from the movement of an almatrino, which has no charge or mass. An almatrino is the smallest initial particle that can fit in our imagination, which began to move by itself.

In the nothingness nothing exists, therefore, God cannot have been in the nothingness to create the Universe. In other words, God does not exist. God is only a philosophical concept that has filled an explanatory void in the minds of most human beings. This theory of God caused a great number of mythologies to be derived; and from mythologies, religions arose; and from religions, the different philosophical currents of how each one perceives God were formed; since, each one of these mythological beliefs, will give us a different concept of God.

As can be seen in Figure 1, the movement of the upward pointing electronic energy 2 and 4 generate a magnetic energy +FB that rotates in a left to right direction. The flow of energy can be demonstrated by the right-hand rule shown in Figure 4.

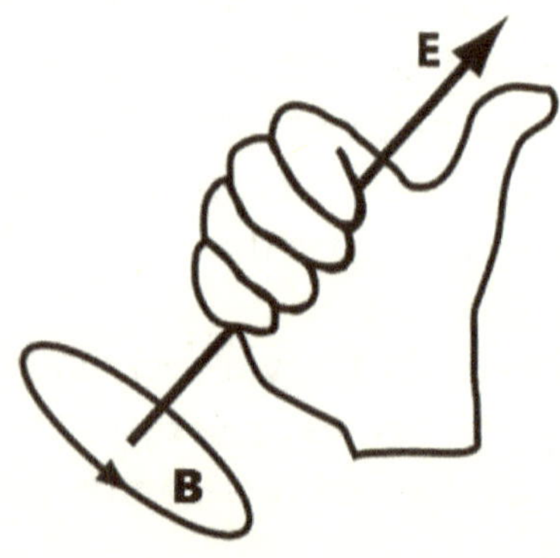

FIGURE 4

MOVEMENT OF ENERGY ACCORDING TO THE RIGHT-HAND MNEMONIC RULE

This integration of the magnetic energies 2 and 4 FB in Figure 1 is equally spontaneous and positive; since, the electronic energy is spinning in the same direction.

It would be like watching two terrestrial tornadoes spinning in the same direction: the two tornadoes would integrate spontaneously and a new tornado spinning in the same direction will be formed. The energy of this new tornado will be the sum of the two tornadoes that merged. The value of the total energy of integration does not change, only the two energies are added together; this will give us the energy of the new tornado that was formed. But if the two tornadoes are spinning in different directions, these two tornadoes would repel each other, and would go off into space on different paths in a disorganised way.

But, in a logical and organised way, an infinite amount of energy levels was formed; whose movement, we can describe by bosons and fermions.

These energy levels are in fact probabilities, which we can describe by means of bosons and fermions, since, as we said, we will not know what is the value of this initial energy to be able to call it quantum energy levels, since energy cannot have fixed values, that is to say, energy cannot have quantum values. In reality, the values defined by quantum energy are probabilities.

So, at that first energy level, we will have the integration probability for two electronic energies rotating in the same direction. For example, for the positive electronic energy the probability is: $[(+1/2) + (+1/2)]$. The positive or upward-pointing electronic energy integrates spontaneously and the positive electronic matter of the electronic nuclei is formed.

Whereas the negative or downward-pointing electronic energy forms the negative electronic matter of the electrons, i.e. $[(-1/2) + (-1/2)]$. This is a form of negative electronic energy. As mentioned, the terms positive and negative do not have a mathematical meaning, they only indicate a direction of spin; and, in the physical sense, this refers to the flow of electronic charges, in order to differentiate between energies spinning in opposite directions.

These forms of energy are already integrated. The two integrated positive electronic energies form the positive electronic matter of the electronic nuclei. Whereas the negative energy when integrated forms the negative electronic matter of the electrons.

These two kinds of positive and negative energy can be further integrated; and for this integration to happen, it has to be spatially. So, electrons will be spatially integrated with nuclei to form electronic atoms. The electronic atoms form the electronic molecules; and the electronic molecules form all the organic and inorganic electronic matter that so far exists in the Universe. The bosons involved in the formation of electronic matter are the gluons.

Electronic matter, whether organic or inorganic, is not aware of its existence, since organic or inorganic matter alone does not form life. In other words, electronic matter cannot spontaneously combine with magnetic mass. For the integration of magnetic mass with electronic matter to happen in a spatial way, the electronic matter has to be evolutionary. The only evolutionary electronic matter is a diploid, which is generated by a process of gestation of two haploids through sexual union.

This energy rotating in the same upward direction produces a magnetic energy rotating from left to right. The positive magnetic energy when integrated forms the intelligent magnetic mass of a positive spirit; that is to say, this is the magnetic mass that is aware of its existence. It is the positive magnetic mass that directs the positive electronic matter of a male being and it is he who carries the haploids that he carries in his gonads.

The masculine being, or that which originates from the integration of positive magnetic energy, can be the spirit of a man, a tiger, a cat, a rooster, a male insect, a whale, a tree or a bull. It will be any being that is made up of the positive magnetic mass that moves the positive electronic matter of a male living being.

In the same way that the positive or left-to-right rotating magnetic energy is integrated, the negative magnetic mass of the right-to-left rotating energy is formed. The integration of this negative magnetic energy forms the magnetic mass of a female spirit; and for integration with the male being, it carries the female haploid in the form of an ovum in the womb. The feminine or negative being; can be, the spirit of a woman, a tigress, a cat, a hen, a female insect, a whale, a bush or a cow. It will be any being that is formed by the negative magnetic mass that moves the negative electronic matter to form a living being with its female characters.

The magnetic mass of spirits and the electronic matter of physical bodies are chiral; which explains why we have two eyes, two hands, two arms; or a brain with two hemispheres. All bodies that exist in the Universe have a left side and a right side. In other words, all living beings are three-dimensional. We owe the explanation of chirality to Dr. Jacobus Henricus van 't Hoff.

These two magnetic energies are already integrated in the form of magnetic mass, so that the magnetic mass of the spirit was formed by the spontaneous integration of two magnetic energies, thus generating the magnetic mass of a female and a male being. Whereas the electronic matter of the physical body was formed from the spontaneous integration of the two electronic energies.

These two kinds of electronic energy in the form of electronic matter and magnetic mass can be further integrated spatially, but without merging. That is to say, this integration will not result in the formation of a single identity. Instead of an integration, one could say that it is an association. As mentioned, this is not an integration, since these two kinds of electronic matter and magnetic mass can only be physically associated in the physical world by means of a diploid. A diploid was formed from two haploids; one haploid was in the gonads of a male being, while the other haploid was formed from a shell of an egg in the female. Whereas a tree can contain in itself both male and female flowers; and to bear the fruits in which the seeds of the offspring go, the tree needs bees for the fertilisation of the flowers.

Thus, the magnetic mass of the spirit is not integrated in a definite way with the electronic matter of the body. Therefore, the two entities can separate when the matter of the physical body completes its evolutionary cycle that gives form to physical life. This process of culmination of the physical evolution of the electronic body is what we call ageing.

There is no death, for neither the electronic matter of the body nor the magnetic mass of the spirit can be annihilated. There is only a process of dissociation between the electronic matter of the physical body that is associated with the magnetic mass of the spirit. After the process of separation, the electronic matter will remain part of the physical body of the Earth. Whereas the electronic mass of the spirit without the electronic matter body will be able to return to its spiritual world. But it will only be able to take with it the experiences of life in its magnetic memory, because the magnetic mass of the spirit does not contain electronic matter. Thus, the magnetic mass of the spirit cannot take anything physical or material from the physical world to the spiritual world.

The evolution of the magnetic mass of the spirit is achieved through knowledge; and this knowledge is what awakens the state of consciousness of the spiritual being. Consciousness is the knowledge of existence; that is, of the right of all beings to exist in the Universe. Consciousness will be dormant in the magnetic memory of the spiritual being. The awakening of consciousness in the spiritual being will create an ecstasy, which will raise the spiritual being to a higher scale of attainment. Through the knowledge gained, it will create in the spirit a feeling for other living beings, who also have a right to exist in this physical world of the earth.

Thus, physical life on earth represents a station in the infinite scale of energetic levels.

Haploids have no memory, so diploids are not aware of their existence. Diploids are made up of evolving electronic matter; whose evolution ends when the process of the evolution of electronic matter ends; which, as we said, is represented by ageing. Therefore, the organic matter of the body only evolves by adjustments of an electronic character without any knowledge of the process. The spirit being is incorporated into the living being when the diploid is reaching the definite evolutionary stage of a baby.

From that moment on, the spirit will take control to direct its exclusive physical body. However, depending on the knowledge that the spirit has attained, the spirit that is incorporated into the physical body of a baby may or may not be aware of its existence. This explains why every day children are born with great genius. But this genius can be erased by the learned habits of most human beings, which depends on the knowledge that the parents of the baby have acquired.

2

EXPERIMENTAL SCIENCE

Unknowing the purpose of existence is only one quality of what the spirit inhabiting the body of a living being must learn. The quality of ignorance has been for all; that is, for scientists and non-scientists. But we must take into account that it is the scientists who are the thinkers who make the greatest contribution to the search for knowledge. Knowledge must be shared with all beings in society who have the desire to learn, for the true knowledge of the spiritual being who lives temporarily in a physical body cannot be hidden and preserved for his or her own use. It would not make sense, for it is necessary to share knowledge in order to promote oneself on a wider scale of spiritual evolution. Selfishness is contrary to knowledge; for selfishness retards the evolutionary progress of the spirit.

We can demonstrate that, throughout history, the human being who wants to progress with his thought and the reason of thought has passed through a series of contradictions, which can only be clarified through experimental science. At the beginning of human history, the contributions that gave us the basis for knowledge were only of a philosophical nature. But philosophical thinking was important, because the philosopher did not have a scientific instrument to prove what he thought with his reasoning ability.

Among these first philosophical thinkers was the Greek philosopher Claudius Ptolemy, but Ptolemy did not have a scientific instrument to prove his philosophy; therefore, the geocentric thought of Claudius Ptolemy was implanted in the mind of the human being who did not make a reasoned analysis in a logical way, but his way of thinking was based on the idea of a mythology.

A time of 1,500 years of philosophy is enough to develop a religious doctrine, which has lasted until today, since many believe that God created Eve from a rib of Adam, or that God said: "Let there be light" and in a magical way light appeared. This, though illogical, was convincing to the unreasonable minded.

Unfortunately, the majority of human beings think in an unreasonable way, and the minority are those who are not convinced by a purely philosophical idea.

It is part of the philosophical but reasoned concept of the great herd of the German philosopher Friedrich Wilhelm Nietzsche. Even in this century of technological progress, this herd is still part of the majority of human beings.

This is what we have to defeat through the science of knowledge, for it is only the knowledge of the origin of the universe that will awaken the lethargic state of consciousness of most human beings.

Friedrich Wilhelm Nietzsche represents a reasonable change between philosophy and experimental science, for Friedrich Nietzsche applies philosophy to society. Most human beings are driven by the habit and custom of what most people do, but we cannot base our knowledge by delegating logic to a theory alone. Whether it is a scientific theory, or whether it has arisen from philosophical analysis; for it is thought that comes first, and that happens before a theory is put forward.

God had to have thought about what the Universe would be like before He created the Universe. So, God is still a philosophical theory; but it is the theory that has lasted the longest in the minds of most human beings; for, it has not yet been possible to prove the existence of God.

Therefore, the philosophical thought of Claudius Ptolemy is the one that has been most successfully implanted in the human mind; that is, the theory that a creator was the one who created the Universe. But, as we have said, this would suppose that the creator existed before the Universe existed; and in the nothing there is nothing; so, in the nothing there cannot exist someone with the intention of creating a Universe, because we would have to search, who was the one who created the creator who created the Universe; and in that retrospective search, we will arrive to the nothing; and there the philosopher's philosophy would dissipate by itself. The creator cannot create himself; and, though the creator is an energy, energy is created by motion, and there was no one there to move the creator who created the creator who created the Universe.

The English philosopher, politician, lawyer and writer Francis Bacon is considered the promoter to mark the difference between the philosophical and the scientific; since, it was Francis Bacon who first proposed the experimental test to corroborate a scientific fact.

FIGURE 5

**THE ENGLISH THINKER, FRANCIS BACON, WAS THE FIRST TO
TO PROPOSE EXPERIMENTAL SCIENCE IN ORDER TO BE ABLE TO TEST
A THEORY**

Francis Bacon proposed that, through scientific reasoning, it was possible to eliminate the preconceived notion of the world. In other words, Francis Bacon argued that it was possible to study the human being by means of experiments. Francis Bacon said that it was possible to study the human being by means of a theory and an experimental science by means of a series of observations. But these observations must be verified by experiment. Francis Bacon said that scientists should above all be suspicious and should not accept explanations that could not be verified by observation and experiment, i.e., as if they were facts that could only be explained in a philosophical way.

It was Francis Bacon who came up with the concept of logic, with the aim of applying and testing reasoning by means of experimental proof. Since, since ancient times, the notion of the origin of the Universe was practiced by means of numbers; that is to say, reaching a conclusion from data that had no support; which cannot be corroborated only with an assumption without an experimental proof.

The experimental method that Francis Bacon proposed represented a breakthrough for the scientific method, as this method was a fundamental reason for the complementation of a scientific hypothesis. However, despite the logic, some scholars of philosophical thought objected to Francis Bacon's experimental science.

Thus, Francis Bacon led a struggle of ideas to object to the unscientific philosophy of Aristotle's philosophical thought, because Aristotle's philosophy limits the progress of applied science. That is to say, philosophy does not allow us to make projections in time. As the French astronomer, physicist and mathematician Pierre-Simon de Laplace later analysed.

FIGURE 6

PIERRE-SIMON LAPLACE SAID: PHILOSOPHY DOES NOT ALLOW US TO MAKE PROJECTIONS IN TIME

For Francis Bacon, if we had followed Aristotle's ideas, these philosophical ideas would have led us down the path of contradictions and the paralysis of knowledge, which is what gives the impetus to human thought.

Francis Bacon criticised Aristotle's philosophical method

because of its practical incompetence, for Aristotle's philosophy was only a symbolic argument, which was useful only for elegant speeches, the colouring of debates and discussions without arguments, but not for the purpose of profit or to produce works that would serve for the improvement of the thinking of the human race.

Francis Bacon refers to Aristotle's philosophy as one that leaves scientific research without any basis, because Aristotle's ideas only revolve around a mythological act. In other words, purely philosophical ideas do not have a reason that gives solid support to human knowledge.

Thus, experimental science is needed in order to verify the origin and nature of what is thought by reasoning. Francis Bacon's proposal is based on proving a philosophical idea by means of an experiment, and it will be with the aim of dominating the force of nature by means of reason that originates in human thought.

But Galileo Galilei appeared with a scientific equipment, and with this instrument, Galileo Galilei was able to overthrow all kinds of philosophical thinking by means of an experiment with a small spyglass. Through his telescope, Galileo Galilei was able to see the vast immensity of the Universe. However, the pope of that time did not want to observe the Universe through Galileo Galilei's spyglass; perhaps, in order not to contradict the philosophical belief of creation; or perhaps, he did not want to see its true creator through Galileo Galilei's telescope.

According to the philosophy of Claudius Ptolemy, the Earth was the centre of the Universe. So, for the Catholic Church everything was clear about the creation of the Earth, or according to the appreciation of Claudius Ptolemy's philosophical thinking.

Aristotle, with his philosophical argument, thought that the creator of the Universe used numbers to do his work; but it is something that does not fit in the logic; or let's say, it did not fit in the logic of Francis Bacon.

It would be the thinker Euclid who would use a compass and a ruler to form geometrical figures and prove Aristotle's ideas of numbers. However, this idea of Euclid's did not convince those who thought like Francis Bacon either. But it was what allowed the birth of an anti-religion different from what the Catholic religion proclaimed. Then, this philosophy would become the doctrine of the creation of the Universe by a projectionist; for, the creator of the Universe did not use only Aristotle's numbers; but, to create the Universe, the creator used a compass and a square; which, produced a contradiction on the religious side as to the creator of the Universe.

On the other hand, the Catholic church would rely on the law of the inquisition to condemn Galileo Galilei along with his thoughts written in a book, in order to erase them by burning them at the stake. However, the Catholic church forgot to burn Galileo Galilei's spyglass; then came the telescopes of the German astronomer William Herschel, the American astronomer Edwin Hubble and the latest one in January 2022, named James Webb, after the American James Edwin Webb.

Unfortunately for the pope, as shown in Figure 4, energy is produced by motion; and the particle that created the Universe started moving by itself and generated the energy by motion. The Universe will not stop moving; that is, the Universe was not created; rather, the Universe is in the process of creation. The Universe will also not stop being created as long as there is motion. But, the motion of the Universe will not end, even if its point of rest is beyond infinity.

Albert Einstein made a mistake, both philosophically and scientifically, when he came up with the theory of relativity. Or let us say, it was the Einstein family to include in this group Mileva Marić; since, the Einstein family, assumed that the magnetic mass is equal to the electronic matter; and that the initial electronic matter of the Universe was imaginary; since, they relied on a mathematical function of Johann Carl Friedrich Gauss and on a Euclidean geometry.

Unless Albert Einstein copied the theory of mathematical relativity from his wife Mileva Marić; and Mileva Marić was a victim of the Matilda effect, because at that time the achievements of women scientists were not recognised. Therefore, the credit for the scientific work went to Mileva Marić's husband, because at that time women did not have access to study together with the barons.

The Matilda effect was described by the feminist Matilda Joslyn Gage in her book "Woman as Inventor". The term Matilda effect is related to the Matthew effect, where the reputation of a well-known scientist is more important than that of an unknown scientist. Often an unknown scientist seeks the support of a well-known scientist to publish his or her work. A work of the unknown scientist, but which can be the most important work to change the way of thinking of the human being.

Even today there is still this classification called social status; for example, where a doctor without notoriety may have more popularity than a renowned scientist. Or the impact of a scientific work on a reader will be different if the author of the work is a woman or if the author is a male scientist. The preponderant character of a male being comes from the positive energy of an almatrino that formed a positive magnetic core.

The Sun is a continuously growing electronic nucleus, because the positive charge accumulates in the electronic nucleus; whereas the planets are actually the negative electronic charge, whose number of negative charges equals the positive charge of the Sun. Thus, the positive electronic charge of the Sun must be equal to the negative charge of its satellites. For example, the negative charge flows from the Earth satellite to the core of the Sun.

Negative electrons can separate from the positive nucleus; therefore, electrons can travel in the form of electromagnetic radiation, generating the separation of photons, and with it, the bright effect of light. It is through radiation in the visible range that we can see planets and objects on planets. This phenomenon of photon separation is what gave rise to light. In the beginning there was no light; the nascent Universe was dark, because there were no planets; therefore, the Universe went through a period of absolute darkness.

Knowledge is like light that illuminates the mind; therefore, the human being will have to go through his period of mental darkness. The arrival of the knowledge that is to be disseminated to minds that think analytically will depend on the time of the work, and this will happen until the new society has mastered scientific ideas, for it is easier to implant a philosophical myth than scientific reasoning.

But, at this moment in human history, through amazon.com and social networks, the arrival of this knowledge can be said to be immediate; but, in addition, the information disseminated will reach millions of people. If knowledge has a reason that manages to impact knowledge, the reason for knowledge will remain part of the evolution of the human being, as reason will always be above knowledge.

However, going back to the search for reason by means of

a scientific instrument; in this case, there may be another conflict that is only because of ignorance, but not because of reason. For example, a telescope will always point towards the chaos of the Universe; because it will be impossible to see from Earth the centre of the Universe. So, no matter how sharp the pictures sent by the James Webb Telescope are, it will not be possible to deduce from these pictures how the chaos of the Universe was formed from an order. Therefore, the most logical thing to do is to start analysing how the Universe was formed from an order that is at the centre of the centre, that is, at the very centre of the Universe.

An order that could only exist in the beginning, considering that the Universe started to form from the motion of an almatrino; and the order of the Universe still exists, but what we can see is a small area with respect to the large size of the Universe; it is this area that we can observe with a telescope; but this small area looks like chaos to us.

To the analysing scientists, the Universe is a mess that will have no end point; for, what we could call the boundary of the void, this edge grows or recedes into nothingness at infinity.

However, the equation $Ev = m_0C^3$, is the one that explains in the most logical and evident way, how the Universe started to form from an order, only by the movement of a minimum amount of energy.

Stephen Hawking realised that, with the experimental science of Francis Bacon, the explanation given by philosophy was seriously flawed. Let us say so, lest we have to say that philosophical thought is dead; for, apparently, the inquisition that sought to erase the reason of Galileo Galilei's logical thought, though it has no bonfire to eliminate the memory of the thinking human being, still the flame of that bonfire lives on in those who refuse to accept the reason of knowledge. They do not want to see the clear photographs that the James Webb telescope is sending us in order not to contradict a doctrine; or they are adapting the doctrine, placing in a convenient order the photographs sent by the James Webb telescope. For which the religious say: behold the greatness of God!

But the trajectory of the Universe and the spirit cannot be modified or extinguished, because they are eternal. It is only the events of electronic matter that change, for these changes will not be repeated; for the Universe moves forward, and it is the combination of electronic matter that changes.

We need to know what this process of the birth of the Universe and the existence of the spirit of all living beings is like, in order to be able to live in harmony at this point in the vast Universe, for the Universe is becoming larger with each passing instant. Not all human beings are aware of their existence and the existence of the Universe, but it was through the energy emanating from the Universe that everything in the Universe came into being.

However, the Universe is being formed by electronic activity; therefore, the Universe is also not conscious of its existence. The conscious part of the Universe is the magnetic mass of the spirits, which are formed from the integration of the magnetic energy emanating from the movement of the electronic energy that creates the movement of the Universe.

In other words, if the Universe were static, no energy of any kind would be generated in the Universe. Therefore, we can say that the Universe evolves by generating electronic and magnetic energies without being aware of their existence. The Universe grows in an exponential and progressive manner, because the Universe is imploding within an absolute vacuum; but this vacuum has no boundary. Thus, in the vacuum within which the Universe is imploding, there are no forces that oppose the accelerated growth of the Universe.

The boundary of the vacuum is infinite; because in the vacuum there is nothing that can stop the growth of the Universe. The Universe will not be able to finish filling with energy the vacuum within which it is imploding, because as the Universe moves, more energy is produced, and the outer edge of the Universe continually expands or recedes to a value where nothing exists. In nothingness there is not even infinity.

So, all spirits must learn how to coexist in the great Universe. But the human being thinks that he or her are the only being that exists in the Universe; therefore, the human being searches in a useless way for a place equal to the Earth to live alone. Or he looks for a planet similar to the Earth, so that this planet can give continuity to the existence of his physical body, because the human being does not yet know that he or she are in reality a spirit that was incorporated into a physical body when the human being's body was a baby in the mother's womb.

But to search for a planet equal to the earth will be an impossible quest, since only the spiritual part of the human being is eternal. So, it makes no sense to travel in physical form to find an earth-like place, because the changing organic matter only exists on earth. The spiritual form of the human being is the only one that can travel faster than the fastest spaceship in existence; however, in order to be able to travel at that great speed, the spirit of the human being will have to separate from his physical body.

On earth, the spirit can only be incorporated to ride on a baby. When the baby is born, the spirit riding in the baby's body will have to start moving at the speed of a baby's physical body. This riding will be progressive as the physical body grows and will end with the ageing of the physical body. The physical body cannot move as fast as the spirit, because if the physical body were to travel at the speed of the spirit, the physical body would become electronic energy.

An embryo was formed from a diploid; and a diploid was formed from two haploids. So, in the Universe, we are not going to be able to find a haploid with a telescope, because it is not logical. A couple consisting of a female and a male of the same species would have to travel in a spaceship; and this couple would have children, grandchildren and great-grand-children in the spaceship, but haploids between families generate genetic modifications.

Haploids are living beings; but haploids have no spirit; because haploids come from the mutation of a virus; and a virus is a transition between electronic energy and magnetic energy. If haploids exist elsewhere, we will not be able to distinguish them; that is, we will not know whether they are the haploids of an animal or of a human being; because haploids are too small to be seen with a telescope; so, haploids can only be seen with a microscope.

Furthermore, the spirit of a human being needs a body to be formed by the combination of two haploids to form a diploid; and this diploid will continue its evolutionary process of replication until it becomes an embryo and then a baby, into which the spirit can be integrated in order to live in a physical body. However, this process is the same for all sexually reproducing beings.

In such a way that the human being is in search of a refuge only for his race but forgets about the other beings that also exist on Earth.

In the human mind there is a debate as to who came first,

the chicken or the egg, but logic tells us that a mutation occurred first. Because it was a mutation, the first bird that formed could not fly with a bladder full of water; so, the bird mutated its physical form. Its reproduction is physical through an egg that is fertilised by a male being. Perhaps a bird was formed by mutation of a manta ray that could fly in the air instead of water. But there are hermaphrodite creatures, such as a salamander that does not need a male salamander to reproduce, the salamander lays the egg by itself without the need of a male salamander.

The bird had to fly in order to nest in trees out of reach of egg-eating predators. In the same way, humans cannot live alone in the Universe. On Earth, for example, a symbiosis was formed between all living species. So, one needs the other in order to exist.

Or let's say that on that journey, which would be impossible to make physically to other galaxies, the human being would have to carry all the seeds and fertilisers to fertilise the soil that will make the grasses grow, to feed his physical body with the seeds that the grasses produce; and enough water to irrigate the grasses. The animals drink water and urinate it; the urea from the animal urine and the ammonia from the fish fertilises the grasses to feed and grow; so, it will be a whole system that the human being would have to carry on his ship in order to move to and survive on another planet.

In the meantime, we will destroy the Earth, even though it has taken the Earth millions of years to form this ecosystem. It is a human mistake, because the unconscious human being does not know that spirits do not need to eat organic substances for nourishment. In order to travel as spirit beings, we do not need to build a spaceship.

So, it is necessary to awaken the state of consciousness of

those who are unaware of their existence and the existence of the Universe, in order to connect all living beings as brothers and sisters, whether they are forming a spirit or are part of a physical body.

On Earth, throughout history, it has been shown that the spirit is unaware of the existence of the matter of the body, since it is impossible for the matter of the body to be aware of the existence of the spirit, which is why the human being assumes that he is only made up of the matter that forms his body.

In this sense, on Earth, the first one we can cite is the French chemist and biologist Antoine-Laurent de Lavoisier; since, Antoine Lavoisier is considered the creator of modern chemistry; therefore, it was Antoine Lavoisier who concluded that:

"...life is a chemical activity".

That is to say that, for Lavoisier, life is only the electronic body made up of matter, which only readjusts or reacts chemically. Antoine Lavoisier became famous because he studied the oxidation of bodies, i.e., the oxidation of electronic matter, but Lavoisier did not consider himself to be the eternal energy of his spirit. Antoine Lavoisier also studied the phenomenon of animal respiration; but it was something that Dr. Ferdinand Perutz concluded with the haemoglobin molecule. However, Dr. Max Perutz was also unaware of his energy as spirit.

Lavoisier studied and analysed air, the law of conservation of mass, caloric theory, combustion and photosynthesis. That is to say, Lavoisier studied everything related to electronic matter; but, without considering himself as the energy that acted to dynamize his physical body. So, Antoine Lavoisier would be the first to establish the knowledge of matter, which led us to study the changes that happen to matter through chemistry, that is, through the science that studies the changes and combination of electronic matter. But this is a theory that we must test, as Francis Bacon said, without taking lightly the result of an assumption of philosophical thought.

The same would happen to other scientists; such as Albert Einstein, Emil Fischer, Max Planck, Paul Dirac or Stephen Hawking; for, if they had had the opportunity to have awakened in time the state of their consciousness or while they were living in a physical body, that life is a duality between the energy of the spirit and the matter of the body, science as the discipline that gives us experimental knowledge, would have led us on the more correct path of reasoning.

Stephen Hawking said when he was diagnosed with multiple sclerosis and the doctors told him he only had a few years to live:

"...I dreamed I was going to be executed; but suddenly I realised there were a lot of things to do, but I would do them, if they gave me an extension".

3

MOVEMENT OF AN ALMATRINO

An energetic particle will always be in motion; therefore, there are no energetic particles that are not in motion. The

Universe is an energy system; therefore, the Universe will always be in constant motion. Before the Universe existed, there was nothing; that is, before the Universe was formed there was no particle without motion. Defined as nothingness, nothingness could not have existed before the Universe was formed; because the Universe began to form at the very instant when the smallest particle we can imagine began to move. Perhaps an instant cannot be defined in the same way that we define time in the physical world, because at that initial moment there was no space to mark what happened between two points.

Therefore, we have had to define the smallest particle without dimensions, without electronic charge; that is to say, without positive or negative charge; without magnetic mass and without electronic matter; since, at that first instant, dimensions did not exist, because at that initial instant the particle was not in motion.

A particle with these characteristics, that is to say, without electronic charge and without any kind of magnetic mass or electronic matter, is what we have had to define for that initial instant as an almatrino.

At an instant, or after the initial point of the Universe, the almatrino was in motion; and the energy of the Universe that was generated by the motion of the almatrino began to be created. The minimum amount of energy which generated and will generate the great energy of the Universe, was produced by the slightest movement of the almatrino in two ways: 1) the positive electronic charge; which, when integrated, formed the positive electronic matter of the electronic nuclei; and the negative electronic charge; which, when this energy is integrated forms the negative electronic matter of the electrons. 2) the positive magnetic charge; which, when united, forms the intelligent mass of a male spirit; and the negative magnetic charge; which, when integrated, forms the negative mass of a female being.

An almatrino represents the smallest particle that can exist; but, in the beginning, an almatrino had no electronic charge, no electronic matter and no magnetic mass, because at that first instant there was no motion and no physical dimensions. Therefore, it is difficult to imagine that someone had to move an almatrino for energy to be created by movement. Being only energy, an almatrino is the smallest particle that we can conceive of by the analytical mind; so, the most logical thing is that the almatrino started moving by itself.

FIGURE 7

WOLFGANG ERNST PAULI DISCOVERED THE NEUTRINO

Perhaps puzzled because he could not find the solution for an energy balance, it happened that, in 1930, the Austrian physicist Wolfgang Ernst Pauli proposed that there should be a particle to compensate for the energy balance, since it was only energy that was missing in the equation of beta radioactive decay. Therefore, such a particle could have neither electronic charge nor mass, since it was only energy that was missing in the energy balance. So, this particle would have to be neutral.

Wolfgang Pauli said:

"...I have done a terrible thing; for I have postulated a particle that cannot be detected".

Wolfgang Pauli therefore called this imaginary particle, which had neither charge nor mass, the neutron.

At the time, the idea of a particle with no electronic charge and no mass could not fit into Pauli's logic, because at that time it was difficult to imagine such a particle. However, because a particle called the neutron already existed, the physicist Enrico Fermi suggested to Wolfgang Pauli that the particle should be called the neutrino, which means small neutron.

Then the Chinese physicist Wang Ganchang proposed the idea of detecting the particle proposed by Pauli from beta decays.

In 1956, experimental physicists Clyde Cowan and Frederick Reines succeeded in developing an experiment to discover the particle. This happened at the P reactor of the Savannah River plant where, on 14 June 1956, Reines and Cowan succeeded in capturing neutrinos.

Clyde Cowan and Frederick Reines sent a telegram to Wolfgang Pauli saying:

"...we are pleased to be able to inform you that we have definitely detected neutrinos from fission fragments by observing the inverse beta decay of protons...".

A neutrino is the smallest particle ever detected by human experiment.

So, we had to define the almatrino as an elementary particle smaller than a neutrino. When I say: 'we have had...', I am referring to us, because I am including you as the reader of this work; since, the mere fact of devoting oneself to read this book, implies a desire to learn and divulge what has been learned, with the aim of awakening the state of consciousness in the human being; and to know, that all living beings are brothers; since, we were all born and will be born from the energy that emanated, that which is emanating and the energy that will emanate from the movement of the Universe.

As for the neutrino, it is estimated that 1×10^{11} neutrinos pass through the area of a thumbnail every second, but despite this enormous number of neutrinos in nature, only a few neutrinos have been detected at underground sites such as the Super-Kamiokande. This is a neutrino observatory located in Japan, and comparatively speaking, this neutrino observatory is the size of a 15-storey building.

Kamiokande was designed to study solar and atmospheric neutrinos, and the decay of protons and neutrinos from supernovae anywhere in our galaxy. The neutrino detector is located 1,000 m underground in the Mozumi mine in the city of Hida in Gifu, Japan. This neutrino detector consists of 50,000 tons of pure water, surrounded by some 11,000 photomultiplier tubes in a cylindrical structure 40 metres high and 40 metres wide.

A neutrino can cross the Earth's globe without being detected. It is estimated that a neutrino can travel the length of a 100-light-year steel rod without being caught.

But if a neutrino is difficult to detect with 11,000 photomultiplier tubes, it will be impossible for us to catch an almatrino. We won't see almatrinos, but almatrinos must be the most abundant energetic particles in the space we don't see.

Almatrinos would fill the dark space of the Universe.

What is known so far is that a neutrino has mass, although the amount of mass of a neutrino is very small.

But, as Wolfgang Pauli did, we have defined an almatrino that has no electronic charge; and no magnetic mass or electronic matter; for, an almatrino is only energy. Energy is an extensive property; that is, energy increases as motion increases. An almatrino generated energy when it began to move.

The electronic energy flowing through the motion of an almatrino will be transformed into electronic mass when the almatrino rotates with high speed. But the rotation speed of an almatrino is greater than the translation speed of the photons of light that form the radiations in the visible range, because electrons have mass.

The boundary that delimits the outer part of the Universe is equal to the boundary of the inner part of nothingness; and this boundary that delimits nothingness with the Universe was moving as the motion of the almatrino was creating space. That event began to happen at the initial moment; but it is still happening with a greater intensity, that is, 13.8 billion years later. The equation describing that speed of motion and initial energy is $Ev=m_0C^3$.

Perhaps, the detection of an almatrino can be achieved in a mathematical way through an extrapolation; since, we will not be able to detect directly or through an experiment an almatrino, because there are no electronic detectors that can capture an almatrino. We will not be able to design an experiment so that the almatrinos leave us a trace, and to be able to demonstrate their existence in a physical way.

The speed of rotation is different from the speed of translation. For example, an almatrino, or any elementary particle, has to rotate, i.e., roll over itself with a high speed, in order to travel in an elliptical orbit. Therefore, the number of spins has to be greater than the speed of rotation in order for the elementary particle to travel through an elliptical orbit. The translational orbit is elliptical, because it is the only way that particles with the same amount of energy do not collide with each other. Thus, the rotational speed will always be greater than the translational speed. The stars are rotating faster than the translation speed by elliptical orbits.

We can calculate the amount of mass m_0 that was formed; that is, what was the amount of mass that was initially produced in the Universe. For this calculation, we will have to start from a minimum time, when we already have two points in the minimum space, i.e., we have a minimum distance and a minimum time. The minimum time is the Max Planck time; that is, $t_{Planck}=5.391 \times 10^{-44}$ seconds, and the minimum Max Planck distance is: $d_{Planck} = 1.616 \times 10^{-35}$ metres.

The comparative experiment would have to be performed with an electron, but the electron is an electron cloud, so the electron has no structure to measure its exact diameter. This is why the electron is defined as a point particle with a point charge, but no spatial extent. If a single electron is observed using a Penning trap, which was constructed by the Dutch experimental physicist Frans Michel Penning, the upper limit of the electron radius can be calculated to be 1×10^{-22} metres. There is a physical constant called the classical electron radius, which has a much larger value of 2.8179×10^{-15} metres. However, the so-called classical electron radius cannot yet be said to have the accuracy of the fundamental structure of an electron.

In such a way that, neither can we know the time and distance below these minimum values, in order to be able to get closer to the initial point or the zero point of the Universe, where the absolute vacuum would be; since, this value has no dimensions; therefore, these quantities below the minimum value do not have a physical sense either. That is to say, if we want to consider quantities smaller than these minimum values, we will not be able to do so; since, below these Max Planck dimensions, space does not exist; so, it has no physical sense, because at that initial point, physical dimensions do not exist.

For quantities equal to or above these minimum Max Planck values, we can calculate the speed of rotation and translation of an electron, whose motion can be represented by the motion of a fermion.

That is to say that, at the beginning, for an electron we can take, to make the comparison with an almatrino, the minimum Max Planck quantity and the classical radius of the electron; that is to say; $2{,}8179 \times 10^{-15}$ metres. Doing the calculations, it will give us, that the rotation value of the electron is $1{,}641 \times 10^{26}$ kilometres per second; while to be able to move through its first elliptical orbit, the translation speed of the electron is $5{,}470 \times 10^{20}$ kilometres per second.

But we are not going to be able to build a Penning trap to study an almatrino. So, comparatively, the rotational speed of an almatrino's energy around itself to move through its elliptical orbit is 300,000 times greater than the translational speed, which makes sense by physical reasoning, since the almatrino had to rotate around itself with a high speed to move through its first elliptical orbit.

But this elliptical orbit became larger and larger as the space created by the movement of the almatrino increased.

Therefore, the rotation speed cannot be infinite. So, this rotational speed has a limit; because at that limit of the rotational speed of an almatrino, the electronic rotational energy will be converted into electronic matter.

The speed at which the electronic energy becomes electronic matter can be calculated from the equation: $v = m_0 C^3 / E$, which is the equation that explains how the Universe was formed from the smallest point by motion.

The relation between electronic energy and magnetic energy is $E = CB$, i.e., there is also a minimum speed at which magnetic energy is converted into magnetic mass, i.e.: $v = mC^2 / E$. In this case, m is the magnetic mass. From this arises the error of Albert Einstein and Mileva Marić.

These are rotational and translational velocities that are above the translational velocity of light. But a speed of light of 300,000 kilometres per second was what Albert Einstein considered, or perhaps it was the Serbian Mileva Marić, to formulate the theory of relativity. Perhaps the theory of relativity was studied by Wolfgang Pauli before Albert Einstein.

However, despite the complexity of theoretical and experimental science, we only know that we can occasionally return to the physical world, as long as we can lose the memory of the previous spiritual being, because the physical memory would not have space to store all the history of our past lives. But to be born remembering past lives would be meaningless, for to incorporate a baby in the womb for the purpose of being born into the physical world is only one more stage in our long process and progress in our spiritual evolution.

As for the physical body, we will not be able to destroy

the electronic matter of the electronic body; for only the electronic matter can be transformed. This electronic matter of the electronic body will only be electronic matter when it no longer has the energy of the mass of the spirit that gave it the form of life.

It will be impossible to disintegrate the mass of the spirit. So, the death of the electronic matter of the physical body and the death of the magnetic mass of the spirit do not exist. It is only a necessary process of disconnection between the electronic matter of the body and the magnetic mass of the spirit. When disconnected, the magnetic mass of the spirit will continue its eternal evolutionary trajectory; whereas the electronic matter of the body will remain on Earth; but it will continue to transform into other kinds of electronic matter.

So, on Earth, only the magnetic energy of the spirit can be temporarily associated with the electronic matter of the body; that is, the energy of the spirit can be integrated with the evolving matter of a baby to form a living being, but without merging as a single identity.

At 5 months after gestation, the evolutionary electronic matter of the future living being will have arrived at an embryo; which emerged from a diploid; and this diploid in turn formed from two haploids. Therefore, until it reached an embryo, the electronic matter of the body of the future being did not have the directionality of the conscious energy of a spirit. Up to this point, the matter of the body changed only through factors of a chemical nature.

At approximately 5 months from the time of gestation, or when the embryo reached the defined conditions for a baby, at that time the magnetic energy of the spirit was incorporated into the electronic body of the baby; and the baby will continue its evolutionary process. So, the spirit attached to the baby will remain in the womb for 4 months, until the birth process takes place.

In that 4-month period, the baby knows its qualities; for, these qualities come with the spiritual being as part of the magnetic memory in the spiritual evolutionary process of each being. The baby before birth already knows that it will be born with a purpose, but it will not remember it in a physical way, as the baby has not formed the hippocampus to store the new memories that correspond to this new stage of life in its long evolutionary process.

The hippocampus will form in the infant's brain during childhood. This stage of childhood ends at the age of 7; and, after that time, the child will not be able to remember his or her time in the womb or how his or her birth happened. Unless their childhood is impacted by an event that is unforgettable.

The foetus has no built-in hippocampus; haploids have no memory, but neither do haploid spirits exist. So, physical memory is something that inherently belongs to the magnetic mass of the spirit; therefore, the spirit brings the knowledge of itself, knows its mission, and the qualities that are proper to its being as a spirit.

Thus, the memory of the spirit can be stored in a physical form in the hippocampus of the child's brain, because the spirit now lives within a physical body, and it is in the hippocampus that the spirit can store the memories of the experiences of this new opportunity for a physical sojourn.

As we said, the memory of the spirit is magnetic; it is an inseparable part of the spirit; therefore, the spirit of a baby can manifest its abilities or skills before it is born, because the baby brings its magnetic memory with it.

After the separation happens, that is, when the spirit separates from the body, the spirit will be able to take its memories with it in an energetic way; which will include in its magnetic memory the form or the stamp of its physical body; that is, the spirit will make an exact copy of the genetic code of its physical body, because the form of its physical body forms an inseparable part of the magnetic memory of the spirit. Therefore, we will be able to recognise the spirit after the spirit has separated from its physical body.

But this process will be the same for all diploids that come from living beings; so, we would have to find out whether the energy that moves the physical body of a plant, a bee, a spider or the body of an ant is not a spirit, but only an energy that makes the body function through movement. Perhaps it is not only an instinct, as bees and ants, for example, have an organisational system that is physically advanced.

Whereas plants have to display their exquisite colours and scents to attract bees, because plants cannot move to have sex, so plants need bees to pollinate them.

We might think that spirits can see the entire electromagnetic spectrum and a more variable mix of colours. But we, from the physical world, will not be able to see what the spectacular world of the spirit world is like. That is why those who remember because they have returned from the spirit world to the physical world say that the spirit world is wonderful.

On Earth, physical life was formed through a biological balance; that is, humans need other animals and trees for shade, which is what allows them to attenuate the radiations of light emanating from the flares of the sun's corona. He also needs plants for nourishment in order to exist in this physical station, which represents the infinite scale of energetic levels.

Solar flares are produced in the solar corona thousands of kilometres from the surface of the Sun. For example, at this time, we do not know what may be happening in the Earth's stratosphere, because we do not see it. We can observe the solar corona flares from Earth, but we cannot see the solar surface from Earth. A spiritual being can live on the surface of the sun. Those who cannot live on the sun are human beings, as long as they are in a physical body made of electronic matter. The physical body of a human being would only volatilise when it wants to cross the solar corona. But the spirit of a human being can live on the surface of the sun.

It is said that the inhabitants of Easter Island destroyed everything on the island. They cut down the trees and could no longer build canoes with the wood from the trees to go fishing. Maybe there, on Easter Island, cannibalism took place. So, they all perished when the last cannibal ate the penultimate cannibal, and the civilisation of the Easter Islanders died out.

Just as it happened to the inhabitants of Easter Island, if they continue as they are going, the inhabitants of Earth will destroy Earth and life on Earth; and the inhabitants of Earth will become the only Easter Islanders in this eternal Universe.

The only ones who can disseminate this information are you as readers of this book, but this is the only way we will be able to change the way human beings think. So, the fact that you have read this book also implies a responsibility. So, you can prepare your lectures with your own discernments, but it will be for the purpose of sharing this information with others. So, that we may succeed in changing the history of humanity; and that the human being may be directed in this coming and going on a path that each day that passes will be better and better.

THE AUTHOR'S WORK

Graduated from the School of Chemistry, Faculty of Sciences, Universidad Central de Venezuela, with a degree in Chemical Technology. Postgraduate studies in Food Science and Technology. Special work on the chemistry of natural products and the chemistry of diseases. Chemical process designer. Books you can locate on Amazon.com®. These books should be subject to revision as we become clearer about how the Universe was formed, so try to read the latest edition of each book. These books are: "The Chemistry of Cancer". "The Chemistry of Diabetes. "The Heart Attack". "Alzheimer's". "The Chemistry of Arthritis". "The Chemistry of Thought". "The Chemistry of the Spirit". "How the Universe was formed". "The Expensalists". "Why You Shouldn't Eat Meat". "The Micro World". "Does God Really Exist?". "Objecting to Albert Einstein's Relativity". "Divining the Future". "The Mistake of the Great Scientists". "Life on the Sun". "The Universe before Zero Time". "The Energy of the Spirit". "The Origin of Cancer". "The World of Cells". "The Chemistry of Disease". "The Particle that Created the Universe". The Chemistry of Cancer, seventh edition. The Chemistry of Diabetes sixth edition; The Chemistry of Heart Attack fourth edition, "The Chemistry of Memory"; The Chemistry of Arthritis third edition. "The Creative Power of the Mind. The Particle that Formed the Universe, third edition. "The Initial Mass of the Universe". "You Shouldn't Eat Meat". "The Origin of the Body and the Spirit". "Worship the Universe". "Sugar an Enemy in the Kitchen". "Time Travel". The Chemistry of Diabetes, Issue 7. The Chemistry of Heart Attack Issue 5. The Memory of the Spirit Issue 1, The Chemistry of Arthritis Issue 5. "The Starting Point of the Universe" The Particle that Created the Universe Issue 5 "The Evolution of Spirit". "The Life of Spirit". "Rewriting Science". "The Beginning of the Universe". "Spiritual Growth". "Coupling of the Spirit with the Body". "The

Origin of Life". "Death Does Not Exist". The Chemistry of Cancer, final edition. The Particle that Created the Universe, final edition.

45